Cambridge Elements ≡

Elements in Grid Energy Storage
edited by
Babu Chalamala
Sandia National Laboratories
Vincent Sprenkle
Pacific Northwest National Laboratory
Imre Gyuk
US Department of Energy
Ralph D. Masiello
Quanta Technology
Raymond Byrne
Sandia National Laboratories

ENERGY STORAGE
ARCHITECTURE

C. Michael Hoff
Hoffpower, LLC

CAMBRIDGE
UNIVERSITY PRESS

University Printing House, Cambridge CB2 8BS, United Kingdom

One Liberty Plaza, 20th Floor, New York, NY 10006, USA

477 Williamstown Road, Port Melbourne, VIC 3207, Australia

314–321, 3rd Floor, Plot 3, Splendor Forum, Jasola District Centre,
New Delhi – 110025, India

103 Penang Road, #05–06/07, Visioncrest Commercial, Singapore 238467

Cambridge University Press is part of the University of Cambridge.

It furthers the University's mission by disseminating knowledge in the pursuit of
education, learning, and research at the highest international levels of excellence.

www.cambridge.org
Information on this title: www.cambridge.org/9781009013932
DOI: 10.1017/9781009028844

© Cambridge University Press & Assessment 2022

First published 2022

A catalogue record for this publication is available from the British Library.

ISBN 978-1-009-01393-2 Paperback
ISSN 2634-9922 (online)
ISSN 2634-9914 (print)

Energy Storage Architecture

Elements in Grid Energy Storage

DOI: 10.1017/9781009028844
First published online: June 2022

C. Michael Hoff
Hoffpower, LLC

Author for correspondence: C. Michael Hoff, cmichael.hoff@gmail.com

Abstract: Energy storage systems (ESS) exist in a wide variety of sizes, shapes, and technologies. An energy storage system's technology (i.e. the fundamental energy storage mechanism) naturally affects its important characteristics including cost, safety, performance, reliability, and longevity. However, while the underlying technology is important, a successful energy storage project relies on a thorough and thoughtful implementation of the technology to meet the project's goals. A successful implementation depends on how well the energy storage system is architected and assembled. The system's architecture can determine its performance and reliability, in concert with or even despite the technology it employs. It is possible for an energy storage system with a good storage technology to perform poorly when implemented with a suboptimal architecture, while other energy storage systems with mediocre storage technologies can perform well when implemented with superior architectures.

Keywords: energy storage, architecture, battery, grid storage, battery management

ISBNs: 9781009013932 (PB), 9781009028844 (OC)
ISSNs: 2634-9922 (online), 2634-9914 (print)

Contents

1 Overview

When we speak of the *architecture* of a system, we refer to how its components are interconnected, assembled, configured, and controlled. The outside of a system may be a flashy, industrial-designed surface with user interfaces, but the inside contains components that are interconnected in such a way as to perform the expected functions for which it was intended. Analogously, the architecture of a building is the design of the essential structure, including beams, walls, floors, and infrastructure, underneath its outer skin. This structure supports the building's functions and the myriad of human's activities as they occupy and traverse its interconnected rooms and hallways. Similarly, the architecture of energy storage affects the flow of energy and matter through a system of interconnected wires and pipes, into and out of vessels or chemical states, while supporting the customer's grid performance and efficiency.

1.1 Architecture Objectives

Ideally, the combination of optimal energy storage technology and architecture will provide the maximum benefit to the customer's grid while maintaining the highest availability and safety, and the minimum amount of lifetime cost for its operators. These are worthy goals, but in reality it is not possible to perfectly achieve every goal with a chosen technology and architecture. Some architectures will enhance a technology's reliability; some will enhance short-term performance, while others may lower the system's costs.

> When choosing the architecture that is most appropriate for a particular project or market, system designers should decide which aspects are most important for their customers, and then select an architecture that promotes those aspects.

The following sections describe some common architectures for the fundamental subsystems of energy storage and indicate how they achieve important application attributes, such as reliability, performance, cost-effectiveness, and safety.

1.2 Energy Storage System Subsystems

Energy storage systems (ESS) are comprised of a set of subsystems that delivers electrical power and energy services to a load or an electric grid while simultaneously ensuring proper working conditions and optimal operation of its components. The four fundamental subsystems of an ESS (depicted in Figure 1.1) are energy storage, power conversion, thermal management, and energy management.

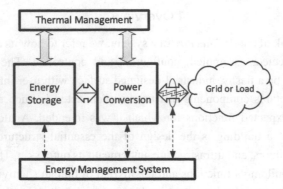

Figure 1.1 A typical arrangement and interaction of the four
ESS subsystems

Energy stored in the ESS is converted to useable power by the power conversion subsystem, which also regulates the flow of stored energy to and from the grid or load. The thermal management subsystem maintains optimal operating temperatures for the ESS components by either adding or removing heat from them. Finally, the energy management subsystem monitors the other ESS subsystems and the grid conditions and controls the operation of each subsystem according to a pre-programmed plan that maximizes its benefits for its owner.

1.3 System Efficiency and Losses

Efficiency is an important aspect of energy storage since it affects the economics of the energy storage project. The more energy lost during the storage process, the more generation is required to compensate for this loss; thus, the more costly it will be to operate not only the energy storage, but the grid as well.

Losses that contribute to the reduction of efficiency can be found in every subcomponent of the system. There are conduction losses due to the resistance of every wire, connector, and switching device from the grid all the way through the last battery cell. Further losses in the batteries stem from electrochemical reluctance of ions to move through the various layers within the battery structure, as well as from other chemical reactions that take place during the storage process. The power conversion system (PCS) has its own set of losses, attributed to its switching devices, as it processes the energy going in and out of storage. The transformers (XFMR) have both conduction losses and standby losses associated with the large magnetic core's magnetization. Most of the subcomponents also have some auxiliary power losses from controls, monitoring, cooling, and heating.

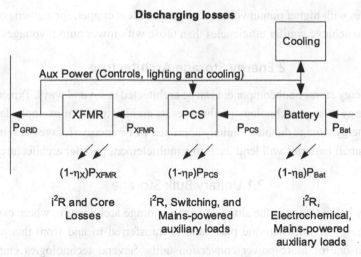

Figure 1.2 Diagram of losses during a discharge of an energy storage system

Figure 1.2 is a useful diagram for calculating the losses during discharge, in each of the elements of an energy storage system.

Looking at the diagram of losses in Figure 1.2, the total losses can be calculated using the following formula:

$$P_{loss} = P_{Bat} \times (1 - \eta_B) + P_{PCS} \times (1 - \eta_P) + P_{XFMR} \times (1 - \eta_X) + P_{Aux},$$
$$(1.1)$$

where P_{Bat} is the ideal power coming from inside the battery. This simplifies to:

$$P_{loss} = P_{Grid} \times \frac{1 - \eta_{Total}}{\eta_{Total}} + \frac{P_{Aux}}{\eta_{Total}},$$
$$(1.2)$$

where P_{Grid} is the power being absorbed by the grid at the point of interconnection, η_{Total} is the product of η_X, η_P, and η_B, and represents the total one-way discharge efficiency, not including the auxiliary power losses. P_{loss} includes a P_{Aux}/η_{Total} component because P_{Aux} is supplied by the battery during a discharge, and since that power goes sequentially through the battery, PCS, and transformer, the $1/\eta_{Total}$ factor is applied.

When comparing system architectures, it is important to consider the differences in the energy losses associated with each option. For example, an architecture requiring fewer controls and monitoring equipment will generally require less auxiliary power. Architectures with fewer, but larger transformers will be more efficient, because larger transformers are generally more efficient.

Batteries with higher output voltages will require less copper, for a given power level, to achieve similar efficiencies than those with lower output voltages.

2 Energy Storage Architecture

The energy storage subcomponent can be architected in several ways. Typically, the energy storage technology predisposes its architecture. For example, large, bulk energy storage dictates a unitary approach while energy storage made up of many small batteries will lend itself to a multielement parallel architecture.

2.1 Unitary Bulk Storage

Unitary bulk storage is the simplest energy storage architecture, where excess grid energy is stored in one place and is transferred to and from this place through one or more power conversion units. Several technologies employ unitary energy storage architecture. For example, pumped-hydro facilities store mechanical energy in one large water reservoir, such as behind a dam or on a mountain [1] (see Figure 2.1).

In a pumped-hydro facility, a turbine converts the potential energy of water stored at a higher elevation to electrical power when the grid needs it. When electrical power is plentiful and cheap, it powers the turbine to pump water uphill from a lower reservoir to the upper reservoir. Another example of unitary bulk storage is a compressed air energy storage (CAES) facility [2]. Although there are many variants of CAES, fundamentally, energy is stored in the form of

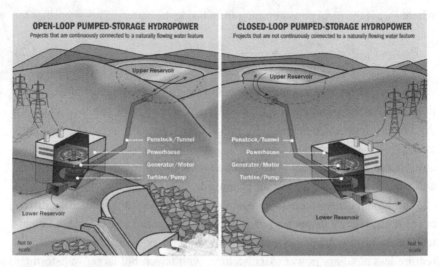

Figure 2.1 Pumped storage facilities showing upper and lower reservoirs connected by a single waterway (Source: US Department of Energy at www.Energy.gov)

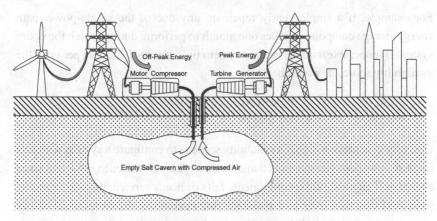

Figure 2.2 Compressed air energy storage system

compressed air in a single large, sealed, sometimes underground chamber (Figure 2.2).

Other examples of unitary energy storage include energy stored in a single tank of liquefied air [3], a tank of chemically reactive material [4], a pool of liquefied salt [5], and even a large pile of concrete blocks [6]. The need for energy storage has been rapidly ramping up and is driving innovation in the industry. Hence, there are many other innovative energy storage means in various stages of research, development, and market readiness.

2.1.1 Unitary Energy Storage with Multiple Power Conversion Paths

If a unitary energy storage facility has only one power conversion path, a failure of a critical component in that path could affect the operation of the whole site. If, for example, the pumped-hydro turbine gets jammed, or the electrical machine throws a bearing, or the electrical transformer overheats, the whole system comes to a screeching halt. Furthermore, prudent, periodic maintenance will force temporary outages throughout the system's service life. This time off detracts from a system's *availability* score.

> *"Availability" is an important characteristic of an energy storage system that defines its ability to operate when it is called upon to do so. The availability score is calculated by dividing the time that the system is fully functional by the time that it is desired to be fully operational.*

$$A = \frac{(Time_{Total} - Time_{Unavailable})}{Time_{Total}} \cdot 100\%. \tag{2.1}$$

For example, if a single yearly repair on any one of the single-power-path energy storage components takes one month to perform, during which the entire system is inoperable, and the desired system up-time is 8,760 hours per year, the availability score, A, would be:

$$A = \frac{(8,760 \text{ hrs/yr} - 30 \text{ days/yr} \times 24 \text{ hrs/day})}{8,760 \text{ hrs/yr}} = 91.8\%. \qquad (2.2)$$

One way to achieve a higher availability score is to configure a system with two or more power conversion units that operate in parallel with each other and sized such that if any one of the components fails or needs service, the other parallel system(s) can take over and maintain full operability. In systems reliability parlance, this is called $N+1$ redundancy, where N is the number of subsystems required for full operation, and 1 refers to the added subsystem that can take over for any one of the N units that may become inoperable in the system. The value of N is chosen to be whatever is practical and cost effective for the application.

Returning to our pumped-hydro case, four waterways can be cut through the mountain to power four sets of turbines and electrical machines as shown in Figure 2.3. If the total power required by the site is P, then each of the power paths could be rated for a power of $P/3$. In most cases, the power in and out of the site will be distributed evenly among the four power paths, up to $P/4$ for each machine. When one of the machines fails, the power will be distributed among the remaining three power paths, up to $P/3$ on each. If by an even

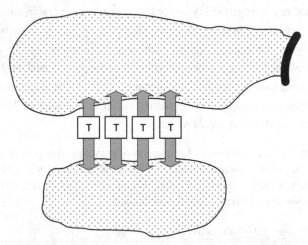

Figure 2.3 Top view of a pumped-hydro facility showing multiple turbines and waterways

smaller probability, another machine fails, the two remaining machines will be left to support the required power. Since they are not rated to handle the full power, P, of the site, the site will be temporarily reduced in power capacity by 33% until one of the broken machines is repaired. While a reduction in full-rated power is not ideal, partial power capability is better than no power capability, as would be the case if only one power path were responsible for the entire site's output.

Therefore, a unitary, single-energy storage, multiparallel power path architecture achieves higher operational availability by providing redundancy in its system components. If any one of the power conversion units goes down, the site will still operate at most or all of its power and energy capacity.

> *The system's availability will be a function of both the probability of individual component failure, the number of power conversion paths, and the full rated "promised" power rating of the site.*

2.2 Multielement Energy Storage Architecture

Not all energy storage consists of large unitary reserves of energy like pumped hydro or CAES do. Increasingly, energy storage is assembled from a number of smaller, discrete energy storage devices, such as battery cells [7], electrolyte tanks [8], or flywheels [9].

When energy storage is manufactured in finite and discrete sizes, the units can be combined in a way that not only scales their power and energy, but also maximizes the performance and availability of the whole ESS. We will explore several different approaches in this section, focusing on battery-based energy storage. However, many of the same approaches can be applied to other storage technologies using the same guiding principles.

2.2.1 Battery-Based Energy Storage

Battery-based energy storage uses discrete-sized components, coupled in a way that provides unified energy storage service to the grid or load. Because they are currently available in relatively small, low-voltage packages, battery cells are connected in series and/or parallel to achieve the appropriate voltage, power, and energy required to provide grid-scale storage services. One might ask:

- How are batteries configured together to achieve a system's goals? and,
- How does one determine whether to connect them in series or in parallel, or a combination of both?

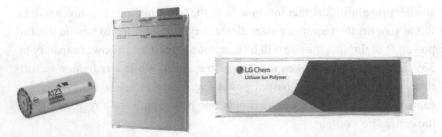

Figure 2.4 Examples of energy storage cells (Source: A123 Systems and LG Chem)

Definitions: In this discussion, a "battery" is a collection of interconnected cells, while a "cell" is the fundamental electrochemical unit of storage. A cell is manufactured having a voltage and charge capacity, both of which vary depending on technology, manufacturer, and product.

An electrochemical cell usually has a terminal voltage between 1 and 5 volts, depending on its chemistry, and can store any number of amp-hours, depending on its size. For example, a lead-acid battery cell's nominal voltage is 2 V, while that of nickel–metal hydride and nickel–cadmium are about 1.2 V. Lithium-ion cells such as those shown in Figure 2.4 nominally deliver between 3.2 and 3.7 V depending on which variant of lithium-ion they are.

Parallel or Series

The approximate energy content (in watt-hour, Wh) of any cell is the product of its capacity (amp-hour, Ah) and its average discharge voltage. To achieve enough voltage, power, and energy for grid services, many cells need to be connected in parallel and/or series. A battery with parallel cells will have a capacity equal to the sum of the individual cells' Ah rating. A battery with series-connected cells will have a voltage equal to the sum of the individual cells' voltage. In either case, the battery will have an energy content equal to the sum of all its connected cells, neglecting efficiency and wiring losses.

To optimally configure the series and parallel arrangement of cells, it is necessary to understand the advantages and limitations of both approaches. Connecting all the cells of an energy storage system in parallel would provide a large amount of current, but at an impractically low voltage. Additionally, since wiring and connector losses are proportional to the square of the current, such a battery would have high operating energy losses. Alternatively, wiring cells in series would result in lower battery currents for an equivalent amount of power; however, there are limits to how many cells can be connected in series. Not only will longer strings increase the monitoring complexity, but it is crucial to keep the total string voltage within the safety limits of the intended PCS equipment and the practical limits of the factory and field handling processes.

Cell Balancing

For both parallel and series configuration, care must be taken to ensure each of the cells shares the system operating power in a manner that achieves maximum benefit with minimal operating cost. In most cases, this means that each cell should charge and discharge at the same *relative* rate with respect to each other, achieving a balanced state of charge among all cells.

When connecting cells in parallel, it is important to ensure that the current entering the parallel bus structure between cells is evenly distributed to each of the cells. This is achieved by careful design of the metal conductors and temperature control of the battery pack. Finite element analysis (FEA) tools can help simulate the direction and volume of current and heat in battery packs with complex geometries, to ensure balanced conditions under all operating modes.

A balanced state of charge will typically exist among cells in series, since the current through each cell is substantially the same. However, over time, there may be some drift caused by slightly mismatched self-discharge rates and charge efficiencies among cells. If this drift is left uncorrected, eventually it will cause enough of an imbalance in the series string that the total capacity of the string will be noticeably reduced. Since the charging process ends when the cell with the highest state of charge (SOC) in the string is fully charged, and the discharging process ends when the cell with the lowest SOC in the string is fully discharged, the maximum achievable capacity of a series string is determined by:

$$Capacity_{achievable} = Capacity_{ideal} \times [1 - (SOC_{max} - SOC_{min})], \tag{2.3}$$

where SOC_{max} and SOC_{max} are numbers between 0 and 1 representing the SOC of the highest and lowest SOC cells in a series string.

To mitigate imbalanced SOC in series-connected strings, many system designs incorporate an electronically controlled cell-balancing mechanism that, over time, gradually corrects for the mismatched self-discharge among the cells.

Sizing Example – Building a Submodule

To demonstrate how a larger energy storage device can be built from smaller cells, let us consider an example of a sizing exercise for a lithium-ion battery. Suppose a cell has an operating voltage range of 3 to 4.15 V and a capacity of 27 Ah. If the nominal cell voltage is 3.7 V, there will be roughly 27 Ah x 3.7 V = 100 Wh of energy in each cell. Larger power and energy can be achieved if we connect a set of

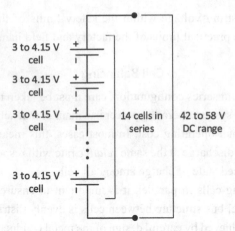

Figure 2.5 Example of a series configuration of battery cells yielding an energy rating of 1400 Wh

these cells in series. For example, as in Figure 2.5, 14 of our 3.7 V cells connected in series will yield an energy rating of 1400 Wh at a nominal voltage of 52 V.

Definition: The arrangement of cells in a parallel or series combination is often called a submodule.

It is important that the cell capacity of each of the series cells be matched as close as possible to each other. Without extraordinary and potentially costly energy-rebalancing means, the smallest capacity series cell will limit the useable energy of the whole string.

Selecting the Battery's Voltage

For the previous example, we chose 52 V as a nominal voltage, but the practical application of energy storage usually dictates the required working voltage of the battery. In general, the working voltage of a battery should be set to match the operating voltage range of the power equipment to which it will connect.

Furthermore, it is helpful to consider the entire operating voltage range of the cells in a real-world application. While the cells are charging and discharging, their terminal voltages vary as a function of their stored energy and rate of change of energy as well as temperature, age, and history. Therefore, any load or power supply to which they connect must be able to operate optimally under this full voltage range. If not, the system designer should adjust the number of cells connected in series to accommodate the connected system's voltage range.

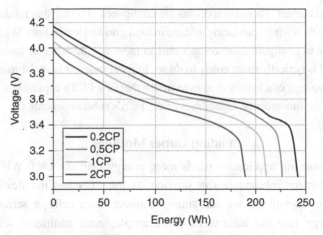

Figure 2.6 Operating voltage range of an example lithium-ion cell while discharging under various rates

Definition: *Figure 2.6 uses the abbreviation CP to differentiate different discharge rates. CP means constant power. 1CP denotes a constant power discharge at a rate where the battery's discharge power is equal to the battery's nominal energy divided by 1 hour. If a battery has a nominal energy rating of 100 Wh, then a 1CP discharge is where the battery is discharging at a constant power of 100 W, a 2CP discharge is where the battery is discharging at a constant power of 200 W, and so on.*

As Figure 2.6 shows, the actual voltage range of a lithium-ion cell varies from 3 to 4.15 V. An n-series string would have a voltage range of n x 3 V to n x 4.15 V. If the maximum rated voltage of a connected load is lower than n x 4.15 V, it will be necessary to reduce the number of cells in series. If this were not done, it would be necessary to operate the battery within a smaller voltage range by limiting the energy put into it. Although it is possible to control the battery's voltage range in this way, it may be undesirable from an economic perspective, since the battery would have to be sized larger than otherwise necessary to compensate for the limited SOC range.

Question: *Why is it that battery string voltage is often set to accommodate the power conversion limitations, instead of power conversion equipment ratings designed around the optimal battery string voltage?*

To answer this satisfactorily, we would need to present an in-depth look at the technologies that make up power conversion equipment today. This Element will not go into too much detail, but to summarize, modern power conversion is composed of discrete components that can tolerate specific levels of operating voltage and current. These components are configured in a way that minimizes the costs of the PCS while interfacing to standardized grid and load points. While it may be possible to configure these components to meet a wide range of battery string voltages, it is typically more costly to do so. In most cases, it is easier to configure the string voltage of a battery system to match that of a PCS's capabilities than it is to change the fundamental capabilities of the PCS's subcomponents.

Building Larger Modules

Now suppose our application needs more energy than the 1400 Wh that our example battery submodule can provide. If the PCS has the flexibility to increase its dc voltage, we can simply connect more cells in series to add more energy into the battery. In our example, each additional series cell would provide an incremental 100 Wh of energy. However, if the maximum voltage constraint imposed by the PCS prevents this, we can add energy without increasing the battery's dc voltage, by connecting additional cells in parallel with the original cells as shown in Figure 2.7.

Notice that in the first of these configurations, we added parallel cells to each of the existing series-connected cells, and in the second configuration, we added a separate string of series cells in parallel with the existing string. The latter configuration will require more monitoring circuits since there are two independent current paths and 14 additional independent cell voltages to monitor.

Other factors to consider regarding the choice between these two configurations are safety and failure recovery, and of course the cost associated with each. For example, when cells are wired in parallel, there exists a possibility that if one of those cells is shorted internally (see Figure 2.8), the other parallel cells will dump a lot of current into the shorted cell and make its failure condition even worse.

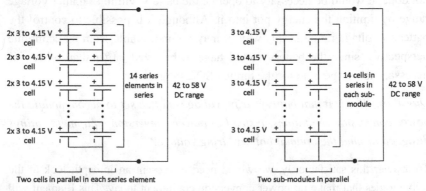

Figure 2.7 Examples of series configurations of paralleled battery cells or battery submodules resulting in 2800 Wh battery configurations

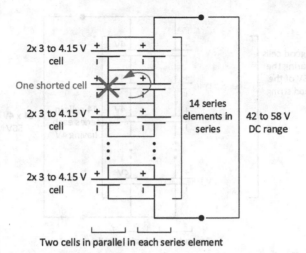

Figure 2.8 Cells in parallel can see dangerous back-feed current when one cell shorts

Note: When lithium-ion cells are wired in parallel, each cell must be protected from "dangerous reverse current flow" that will result if there is a short circuit inside a cell. This can be done with a diode, or more economically with an appropriately rated fuse on the terminals of each cell. This is required for compliance with United States Department of Transportation regulations (see 173.185 in [10]).

In another example, when a string of cells is connected in parallel to another string, there exists the possibility that if one of the cells in one string has an internal short, the other cells in that string will experience an over-voltage condition, because the voltage across the good string will be imposed on the bad string, resulting in the good cells of the bad string having a higher average voltage than that across the good cells of the good string (see Figure 2.9).

To prevent the danger presented by this condition, system designers must implement a means to disconnect the two strings from each other by a controlled switch, contactor, relay, or transistor when such a condition is detected.

Because of the extra cost of the controlling and monitoring electronics in the parallel string configuration, it can be more expensive than adding cells in parallel to achieve higher energy levels. However, this may come at a cost of redundancy and flexibility, which we will address in later sections.

Management and Protection

An arrangement of unprotected and unmonitored cells is not a complete energy storage subcomponent. At the very least, cells should be protected from abusive conditions such as excessive current, internal and external short circuits, and physical abuse. Most chemistries also need to be protected from overcharge,

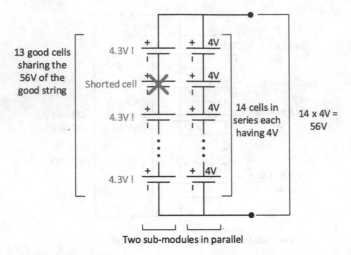

Figure 2.9 Parallel strings experiencing a shorted cell could result in multiple overcharged cells

over discharge, and adverse temperatures. In addition to active protection, the battery's voltage, current, and operating temperatures should be monitored and reported to upstream control systems that can actively keep the battery operating in its most optimal condition. Moreover, the individual cells of some chemistries (notably lithium-ion) *must* be monitored to allow the control system to take steps to protect them.

For many years, ESSs using lead-acid batteries employed very little monitoring and management. Typically, an uninterruptable power supply (UPS) system consisted solely of lead-acid battery cells or monoblocs [11] (some examples shown in Figure 2.10) connected in series and then connected to the PCS through a protective fuse. In fact, some telecom system operators did not even employ a fuse, because the possibility of a fuse failure would cause the battery to stop delivering energy to their critical loads. Lead-acid batteries have the ability to self-balance their charge voltages over extended float-charging periods, and to accept a limited amount of overcharge without catastrophic failure. This is why it is possible to connect several of these cells and groups of cells in series with no autonomous monitoring, to achieve a simple, low-cost energy storage system.

The Case for Protection

Batteries based on other chemistries, including lithium-ion, do not have the ability to absorb overcharge current and cannot balance their charge voltage with the other cells in series. It is well known that serious damage can result

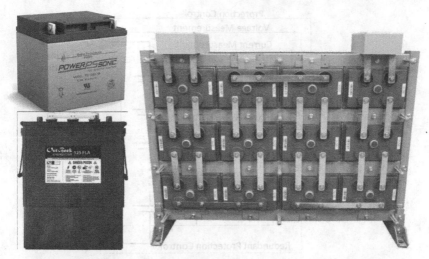

Figure 2.10 Examples of lead-acid monoblocs and UPS
strings (Powersonic and Outback)

from overcharging a lithium-ion cell [12] [13]. Furthermore, lithium-ion cells can react violently to overcharging, overcurrents, overtemperatures, and physical abuse. Hence, most modern lithium-ion battery system designs include monitoring, reporting, and integrated protection to keep them and their users safe. Best-in-class designs incorporate redundant protective mechanisms, further reducing the probability of catastrophic side effects of a failed battery or cell.

Figure 2.11 depicts a typical connection of cells with monitoring and protection circuitry. This diagram shows a configuration of cells, in series and parallel, being monitored by a battery management system (BMS). The BMS also measures temperature, current, and module voltage, and controls switches that can actively protect the batteries from externally applied electrical abuse, if necessary. The diagram also shows that the BMS has a communication link to another control system, which can be another module, or an upstream controlling device. Overcurrent protection is shown as a fuse in this diagram, which is usually the most economical choice of system designers.

Definition*: A configuration of cells combined with monitoring and/or built-in protection is often called a "battery module."(Examples are shown in Figure 2.13.)*

Protection with Monitoring

The module's BMS monitors the conditions of its cells and, when necessary, takes action to mitigate or prevent damage to them. For example, the BMS

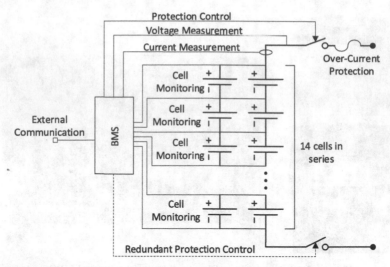

Figure 2.11 A configuration of interconnected cells with
protection features

observes the relative SOC of each of the cells and engages circuits to keep them
in balance with each other. It may do this by dissipating the energy of the cells
with higher SOC, or by moving energy from the high SOC cells to the low SOC
cells using active-balancing circuitry. Additionally, the BMS can warn of
conditions exceeding the cell's recommended limits through external communication to an external controller. If the external controller has control over the
current applied to the battery terminals, it may be able to bring the cells back
into a safe operating state before damage to them occurs.

For example, consider the case where the external controller is a battery
charger and is currently charging the battery module. If the BMS detects that
one of the cells has a high SOC, it can signal the charger to reduce or stop the
charging current while the BMS's balancer circuit brings the high SOC cell
down to match the SOC of the other cells. In this scenario, the BMS thwarted
a potentially unsafe condition by monitoring the cell's state, signaling the
external device, and taking internal action to mitigate the unbalanced SOC
condition.

Independent Active Protection

Additionally, a module's BMS with integrated protection can have start–
stop control of the current entering and leaving the module by controlling
the state of one or more series switches. These switches could be con-
tactors, relays, or solid-state transistors. If signaling to an external controller does not improve a potentially harmful condition, the BMS can open

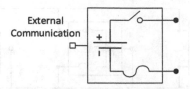

Figure 2.12 Simplified depiction of battery module with connected cells and integrated protection

Figure 2.13 Examples of commercially available 12, 24, and 48 Vdc modules (A123, Delta, Saft, and Simpliphi) with integrated protection

these switches to stop current from passing through the cells and avert a dangerous condition.

The fully protected module can be used as a building block to build larger energy storage systems and is depicted with the symbol shown in Figure 2.12.

Scaling Up – Building the Components of a Larger Grid-Scale ESS

To build large battery systems, such as for grid applications, one must combine modules in parallel or series or both. If, for an intended application, the voltage range of a single module is adequate, then to achieve more energy we can connect the modules in parallel and monitor them with a communication-aggregating device as shown in Figure 2.14.

In this configuration, each of the modules can protect themselves from an internal fault or harmful, externally applied conditions. The master communication device can not only aggregate information from every module but can also take action to coordinate graceful shut-down and start-up sequences among the connected modules.

In the case of a grid-tied ESS, a nominal dc voltage of 52 V is arguably too low. A grid-like 1 MW load on a 52 V battery would result in a current of about 20,000 amps! Even a resistance of a micro-ohm would cause hundreds of watts of power loss. Consequently, grid-tied PCSs are typically configured with dc

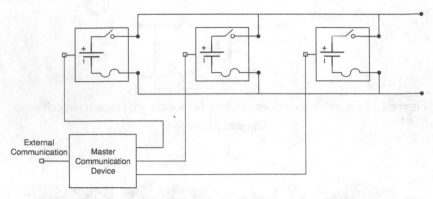

Figure 2.14 Battery modules wired in parallel for more energy

voltages higher than 500 V. For example, if 17 of the 52 V modules are connected in series, the system will have approximately 880 Vdc, as shown in Figure 2.15.

For this configuration to operate safely, each of the modules' protection devices (both active and passive interrupting devices) would have to be rated to handle the whole series voltage of the string. While this may be possible, it may not be cost effective. To save cost, the active protection components of the modules in this string could be consolidated into one BMS, leaving the individual modules with only monitoring and reporting capabilities (symbol shown in Figure 2.16).

When integrating these unprotected modules into an energy storage system, it is vitally important that a central BMS be responsible for the protection of all the modules. This can be done by monitoring the status of each module by communicating to each of the module BMSs, reporting their condition to an external upstream controller, and interrupting the current that is passing through the modules when it detects a critical problem.

Each of the modules should still contain overcurrent protection, usually in the form of a fuse, which is rated for the module's voltage and current. This is important for safe handling and transportation (see 173.185 in [10]) of the module. They must also contain monitoring circuits to monitor the status of the cells in the module as appropriate for the specific technology. Note the "Main Pack Fuse" situated roughly in the middle of the string of modules in Figure 2.17. This fuse gives the entire string of modules further protection from overcurrent and short circuits. It should be rated for the full string voltage and current and be specified to blow *before* the individual module fuses would have otherwise opened in response to an external short circuit.

Definitions: *When a string of modules is connected with protection and communication, it is often referred to as a "battery pack." This storage component*

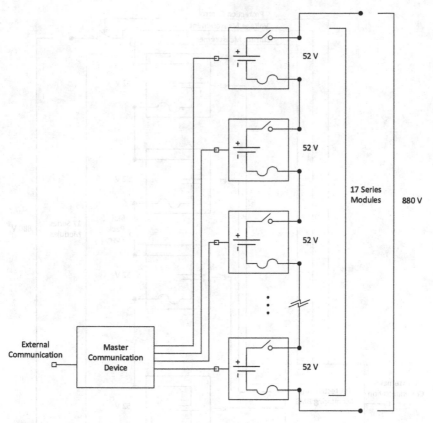

Figure 2.15 Seventeen series modules to yield a grid-scale voltage range of 700–1,000 Vdc

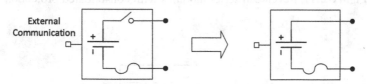

Figure 2.16 Remove integrated protection from modules when not needed

is the most common configuration of an electric vehicle (EV) battery. In grid applications, the modules are physically configured vertically in a cabinet commonly referred to as a "battery rack."

The building block in Figure 2.17 can be further consolidated and is depicted in Figure 2.18. Notice that the main pack fuse is positioned halfway between two half-strings of batteries. There are other fuse positions that deliver the same

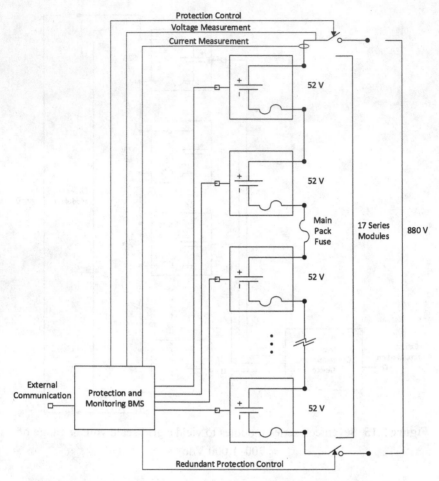

Figure 2.17 Seventeen series modules with consolidated protection

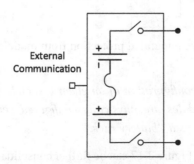

Figure 2.18 Battery rack – grid-scale battery building block

Figure 2.19 Examples of commercially available battery racks for grid
applications (Delta, NEC, LG Chem, and Samsung)

short-circuit protection, but an advantage of this positioning is that if the fuse is
ever blown open, the remaining voltages in the battery rack are about half the
full battery's string voltage, leaving a safer environment for recovery and repair
efforts. Examples of battery racks are shown in Figure 2.19.

2.2.2 Parallel System Architecture

Section 2.2.1 introduced a building block with which we can build larger
systems to connect to the grid. If these blocks have an identical dc output
voltage that is compatible with the grid-connected power conversion equipment, they can be connected in parallel at the system level, as shown in
Figure 2.20, to achieve even greater power and energy.

For parallel systems of batteries (one example shown in Figure 2.21), the bus
connections between the battery racks and the service disconnect switch need to
be carefully designed to not only handle the continuous rated current, but to
handle fault currents and enable each of the racks to discharge and charge at an
equal rate. If the latter aspect is ignored, the imbalanced currents will cause

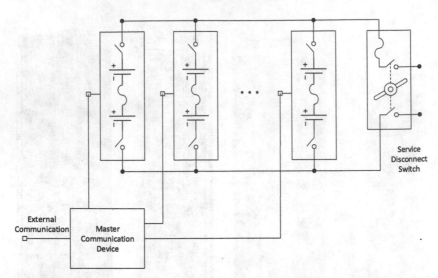

Figure 2.20 Grid battery system (GBS) – parallel arrangement of energy storage racks

Note the "service disconnect switch" with fused overcurrent protection in the grid battery system (GBS) as depicted in Figure 2.20. This is a useful device in situations where the GBS needs to be isolated from other parts of the energy storage system during servicing and transportation.

some of the racks to wear out sooner than others, thus causing the whole system to operate at a reduced efficiency.

The "master communication device" is an additional layer of monitoring that aggregates the information from each of the parallel-connected energy storage racks and may coordinate the connection and disconnection of each rack to the common dc bus. In such an architecture, it is possible for the master communication device to control when each energy storage device connects and disconnects to the dc bus to maintain the health of the individual energy storage devices or increase the collective performance of the whole group.

Balancing Power among Multiple Parallel Energy Units

When energy storage units are connected in parallel with no *active* means of equalizing their shared power, their relative internal impedances will determine the balance of current between each of the units. Variations in impedances can result from variations in manufacturing tolerances, age, operating temperature, performance degradation, connection quality, and wire resistance variations.

Figure 2.21 Example of a group of parallel-connected racks coupled to a dc disconnect switch and master controller (Source: NEC Energy Solutions)

Some of these factors can be mitigated using thorough design and manufacturing controls, while others may be affected by product servicing and maintenance operations.

It is important to control the energy flow from each of the energy storage units such that they contribute their power in proportion to their individual capability or their present condition. Any imbalance in the rates of power delivery will manifest itself as a reduction in overall useable energy, due to decreased efficiency or suboptimal delivery of power. The imbalance will also prevent the group of racks from achieving its fully rated power. For instance, if one or more lagging racks were contributing less current than the others, then to achieve the group's maximum power capability, one or more of the other racks would need to operate above their specified power rating to compensate for the lagging racks. Therefore, if there is expected to be mismatched power flow among racks, *both* the power and energy ratings of the group of racks would have to be de-rated accordingly.

Protecting High Current Carrying Wires and Systems

A dc current is notoriously difficult to interrupt in short-circuit events. Inductance from wiring and loads cause current to continue to flow after an attempted interruption, even pushing through air gaps in the form of a high voltage spark. Thus, manufacturers rate their short-circuit protection devices for their ability to interrupt current and the voltage that drives that current. Grid-scale batteries are particularly challenging because they can contribute many tens of thousands of amps each. Thus, when paralleled together, it is feasible for a configuration of battery units to provide hundreds of thousands of amps into a short circuit.

Aside from providing the ability to carry and interrupt short-circuit currents, wires and power devices need to be sized large enough to survive the short-circuit events without presenting a hazard to operators and bystanders.

> **Sidebar Topic**: *Overcurrent Protection Coordination*
>
> This brings us to an often overlooked, but crucial topic to understand when successfully designing a large battery system. Overcurrent protection, comprised of a combination of fuses, circuit breakers, and actively controlled switches, prevents catastrophic events caused by accidental short circuits both within and outside the energy storage system. Nevertheless, there are more complex aspects of protective devices than simply their rated trip currents. Fuses and circuit breakers also have ratings that determine the voltage at which they will successfully interrupt current, their maximum interrupting current, their interruption speed, and the currents at which partial melting can occur. All these characteristics must be considered when implementing a system with high power and potentially tremendous short-circuit currents. In addition, one must observe the conditions the battery will be subjected to before, during, and after the opening of the protective devices. Special attention must be paid to the order in which protective devices will open in response to various internal or external faults; otherwise, unexpected collateral damage can occur to the battery during a fault.

Using DC/DC Converters to Enhance Parallel Energy Storage Performance

To counter some of these challenges, some manufacturers insert buffering power converters between each energy storage unit and the dc bus. This converter is a regulated dc/dc conversion system which adjusts the relative impedance between its inputs and outputs so that the individual energy storage unit's power levels can be adjusted to match the needs of that unit or to accomplish some other system-level goal.

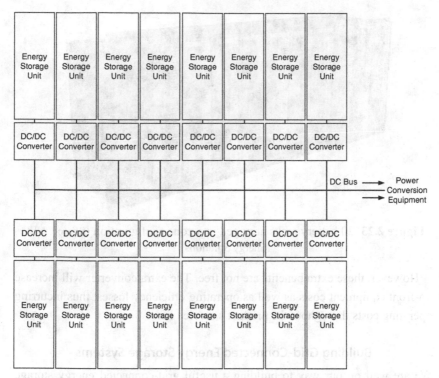

Figure 2.22 Energy storage system with controlled power conversion between the dc energy storage and the common connection bus

The configuration in Figure 2.22 solves some of the issues associated with raw, parallel connections. The dc/dc converter can be controlled to draw an appropriate amount of power from each of the energy storage units, regardless of their potentially mismatched impedances, battery types, rack voltages, or other factors. It also controls the absolute maximum current, even in a short-circuit condition.

In addition, the converters allow new racks to be placed in parallel with older racks. Otherwise, if newer, low-resistance racks are placed in a hard connection with older, higher-resistance racks, the newer racks would discharge and charge faster during a cycle than each of their older neighbors. This causes an imbalance in current and even SOC among the racks; possibly to the point of exceeding individual rack current limits. Systems with an individual converter on each of the parallel strings can achieve a truly balanced SOC condition by adjusting the current from each of the racks to compensate for mismatched ages and other factors.

Figure 2.23 20 battery racks arranged in a zone configuration (Source: NEC Energy Solutions)

However, these extra benefits are not free. The extra converter will increase up-front equipment costs as well as operating efficiency losses, thus incurring operating costs throughout the product's service life.

Building Grid-Connected Energy Storage Systems

We are well on our way to building a useful grid-connected energy storage system. If we have a collection of 20 racks, each with 100 kWh of energy storage, and we connect them in parallel as described in Section 2.2.2 and shown in Figure 2.23, we could provide 2 MWh worth of energy services to a grid through a PCS.

These collections can be housed in a container (examples shown in Figure 2.24), a specially built enclosure or building, for physical protection (more on this in Section 4), to create a complete grid battery system.

Typically, one or more of these grid battery systems would then be connected to a grid-scale PCS to comprise the essential elements of a grid energy storage system (GESS). The one-line representation of such a connection is shown in Figure 2.25.

In fact, multiples of these configurations can be connected in parallel to the grid as in Figure 2.26 to scale the amount of power and energy needed for that site and gain an increased redundancy as well.

If one of the components in a battery system fails, that component can be automatically disconnected from the rest of the battery system, allowing the battery system to operate, but at a slightly lower power and energy capability. If something happens to an entire battery system or its associated PCS, it can be taken out of service by disconnecting the PCS from the grid, allowing the other

Figure 2.24 Examples of commercially available fully enclosed grid battery systems (Saft, Delta)

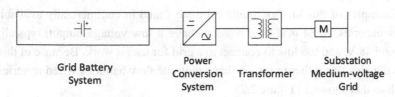

Grid Battery System   Power Conversion System   Transformer   Substation Medium-voltage Grid

Figure 2.25 One-line diagram showing a grid battery system connected to a utility grid

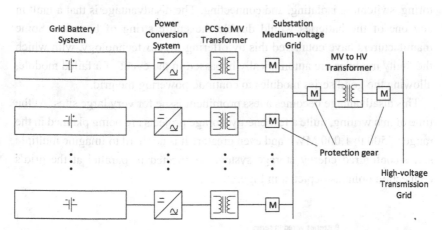

Figure 2.26 Multiple parallel grid battery systems connected to the grid through one point of interconnect (POI)

parallel-connected PCSs and battery systems to continue operating. This parallel architecture offers a high level of redundancy and site availability.

2.2.3 Series Architecture

The paralleled battery concepts described in Section 2.2.2 have advantages in terms of redundancy and scalability, but they come at the price of additional cost. Each of the paralleled components requires separate monitoring, protection, wiring, connectors, and disconnection devices. If the individual energy storage devices were built with larger amounts of storage, and wired in series instead of parallel, as in Figure 2.27, the energy storage system would have less monitoring, wiring, connectors, and disconnection devices.

Of course, the protection and disconnection devices would have to be proportionally larger than those of the parallel architecture, but one could expect the net relative costs to be lower for the larger, but consolidated components.

Examples of this kind of architecture are found in commercially available flow batteries. Most grid flow batteries have a low voltage output, typically around 48 V, and too low to connect to a grid for useful work. Because of this, installations using these products have multiple flow batteries wired in series, such as that shown in Figure 2.28.

In such an architecture, dc current flows in series through each of the flow batteries as they collectively power the grid-connected PCS. The advantage is that there are potentially fewer components performing the functions of monitoring, switching, isolating, and connecting. The disadvantage is that a fault in any one of the batteries would disable the entire string of batteries. Some manufacturers have countered this by offering *bypass* technology, with which the faulty modules are automatically bypassed in the event of a failed module, allowing the good series modules to continue powering the grid.

This disadvantage becomes a less prominent issue for very large sites. At the time of this writing, quite a few energy storage projects are being planned in the range of 500 to 1,000 MWh and even greater! It is not hard to imagine multiple series-configured energy storage systems connected in parallel at the grid's connection point as depicted in Figure 2.29.

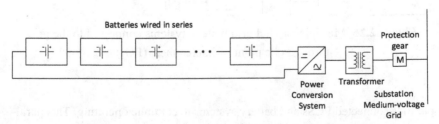

Figure 2.27 Large energy storage units wired in series

Figure 2.28 A commercially available flow battery made up of individual modules that are wired in series to build up an appropriately high enough voltage to power a grid-connected PCS (Source: UniEnergy Technologies)

2.2.4 Reliability of Parallel and Series Architectures

The reliability of any system depends on the probability of failure of its components, the configuration of those components, the system's parallel redundancy, and the ability of the control system to properly adapt to the system's state of operability. There are many books, papers, and even college courses devoted to the study of reliability. The scope of this section is to cover just the basic differences between the reliability of series and parallel battery strings.

Fundamentally, a battery is specified to deliver a certain amount of voltage and current during a discharge. To determine the reliability of a battery, we look at the ability of the battery to deliver its rated voltage and current when one of its subcomponents fails.

In a series-connected battery string, if one of the series batteries fails open (see Figure 2.30), the string's voltage will drop to zero.

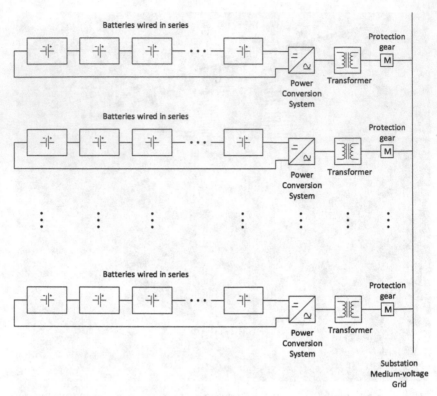

Figure 2.29 Parallel connection of multiple series-configured energy storage systems on the grid

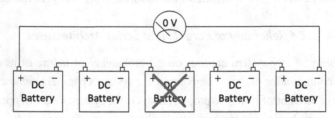

Figure 2.30 Series battery string with a battery failed open

The probability of the failure of the series string is compounded by how many batteries are in series. Written in terms of reliability, R of the string is the product of the reliabilities of N-series batteries:

$$R(\text{system}) = \prod_{n=1}^{N} R_n (\text{series components}). \qquad (2.4)$$

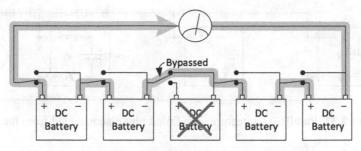

Figure 2.31 Series battery string with a bypassed failed battery

Therefore, the reliability of a series-connected battery will often be less than the reliability of the individual battery. For example, the reliability of 10 series batteries, each having 99% reliability will be $0.99^{10} = 90\%$.

If, however, the individual batteries have a complex fail-over mechanism, which allows the string to continue to operate by bypassing the failed unit (see Figure 2.31), the string can deliver part of its rated voltage, power, and energy. In this case, if an N-series string were oversized in voltage, power, and energy by a $(N+1)/N$ factor, the string will still perform according to its desired specification.

Bypass technology achieves redundancy for series strings but is difficult because of the limited voltage range of the power conversion equipment. Increasing redundancy by adding more series batteries increases the overall string voltage, possibly beyond the upper limits of the intended power conversion equipment. On the other hand, bypassing too many batteries in a string will result in a serious deficiency of voltage. Therefore, this approach has merit only when the intended power conversion equipment has a wide enough dc operating voltage range to accommodate the worst-case scenarios.

It is simpler to increase system redundancy with a parallel architecture, because having more energy storage units in parallel increases the system's reliability, power, and energy without changing the system's voltage rating. A failed storage unit can be simply disconnected from the other units to allow the good ones to continue to operate (see Figure 2.32).

If more units are present in the system than necessary, the system can tolerate a failure of one or more units if the remaining units can support the power and energy requirements of the site's application. Assuming the control system can proficiently disable failed units without affecting the remaining working units, the reliability, R, of the system would be:

Grid Energy Storage

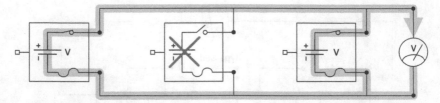

Figure 2.32 Parallel architecture with failed unit disconnected from the rest

$$R = \sum_{n=k}^{N} \frac{N!}{n!(N-n)!} r^n (1-r)^{N-n}, \qquad (2.5)$$

where N is the total number of storage units in parallel, k is the number of storage units that need to operate to meet the site's requirements, and r is the reliability of each of the individual storage units.

For example, if a site needs 10 operating storage units, each with a reliability of 99%, to achieve its power and energy requirements, adding one in parallel to these 10, will yield a site reliability of:

$$R = \frac{11!}{10!(11-10)!} 0.99^{10} (1-0.99)^{11-10} + \frac{11!}{11!(0)!} 0.99^{11} (1-0.99)^0 = 0.995.$$

Adding more parallel units will further increase the reliability.

3 Power Conversion Architecture

Direct current (dc) provided by batteries is not directly useful for grid purposes until it is converted to regulated alternating current (ac) by a power converter. An energy storage system's PCS converts dc of the battery to ac of the grid and regulates the flow of energy to and from each side.

Before 1960, it was common for a pair of dc and ac electromechanical machines, coupled by a mechanical connection, to perform dc to ac conversion (see Figure 3.1). A battery would power the dc machine, acting like a motor, which would mechanically drive the shaft of an ac machine, acting as an alternator (generator that outputs ac), feeding ac power into a load or the grid. The fields of the machines would be controlled to regulate the amount of energy flowing between the dc and ac busses, making it possible to charge the battery from the ac grid or power the grid from the dc battery. The disadvantage of this method is that the machines were large, noisy, and required frequent maintenance.

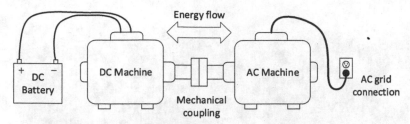

Figure 3.1 Electromechanical dc to ac power conversion

3.1 Electronic DC to AC Conversion

Today, the majority of dc to ac power conversion (one example in Figure 3.2) is performed by electronic switching devices, which are connected and controlled in a way that precisely regulates the energy flow between a dc energy source and the ac grid. Electronic switches, such as Insulated Gate Bipolar Transistors (IGBTs), Field Effect Transistors (FETs), and Silicon Controlled Rectifiers (SCRs) are the most common devices used in modern PCSs. These devices allow for a more efficient, compact, and reliable conversion.

Many texts and articles cover this topic in greater detail, but for the purposes of this Element, the electronic switches in a PCS rapidly turn on and off connections between the ac and dc ports, such that the average time spent during connection states results in the desired average rate of flow of energy. The process of turning switches on and off to achieve a desired voltage or current outcome is called pulse-width-modulation (PWM). The sharp, abrupt PWM current is smoothed out by passive inductors and capacitors before being applied to the dc and ac ports (see a simplified diagram of such a converter in Figure 3.3).

3.1.1 PCS Efficiency and Losses

As mentioned in Section 1.4, it is important to know how much energy is lost from inefficiencies and where the losses originate. A PCS has several sources of inefficiencies. First, the switching devices and the power-carrying device in the PCS are not perfect conductors. They produce heat at a rate equal to the square of the current flowing through them. This is important because it tells us that for a given operating power, a higher operating voltage will result in lower current, lower *conduction losses,* and higher efficiencies. Second, the switching devices do not switch instantaneously, and therefore, incur some *switching losses* every time they turn on and off. This loss is proportional to the switching time, switching frequency and the switched current and voltages. Unlike with conduction losses, higher operating voltages do result in higher switching losses.

Figure 3.2 Modern PCS from EPC Power

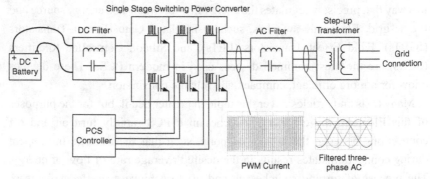

Figure 3.3 Basic components of a modern PWM PCS

Depending on the PCS architecture, this loss can be significantly larger than the conduction losses and should be carefully considered.

The heat generated by switching and conduction losses in the PCS can amount to anywhere between 1 and 3% of the full power rating of the PCS. For a 1 MW PCS, this means up to 30 kW of continuous power must be dissipated and removed from the PCS to keep it operating at its optimal temperature. Depending on the specific PCS design, temperature regulation can be accomplished by a water loop in a cold plate, bringing heated water from the inner core of the PCS to a radiator, or some other cooling device on the outside of the PCS. Other cooling designs may use fans to blow cool air across a set of heat sinks, which are attached to the heat-generating PCS components. This cooling system also uses energy to power fans, pumps, compressors, and all other components of the cooling system.

Finally, every PCS has a control system that monitors and controls the functions of the PCS to make it do what an external controller is telling it to do. The power to run this control system is constant and independent of the power being converted. Typically, the control losses are small compared to all other losses in the PCS, but they should be considered when diligently modeling the performance of an energy storage system. When considering all the losses in a PCS, it is important to understand that they are dependent on the operating conditions of the PCS and are not a fixed quantity or percentage of its rating.

3.2 Electronic DC to DC Conversion

Although, most energy storage is connected directly to an ac power grid, there are increasingly more projects requiring a connection to a dc network instead. This dc network could be a dc microgrid, or more commonly, it could be the dc connection to a solar photovoltaic (PV) array.

In a configuration such as depicted in Figure 3.4, if the sun is bright and the utility load is light, the excess solar energy can be directed into the energy storage system. When the sun goes down and the utility load starts to increase in the early evening hours, the energy storage can discharge its energy into the ac grid.

The advantages of this scheme are:

1. When the energy storage is mostly used to store excess day-time solar energy, the round-trip efficiency will be higher than with an ac-connected grid storage system doing the same.
2. Many power purchase agreements (PPAs) with the local utility, restrict the total ratings of the power conversion connected to a specific site. This scheme allows a fully rated energy storage asset to operate along with an equally sized PV array, behind a single PCS, while still getting optimal use from both.

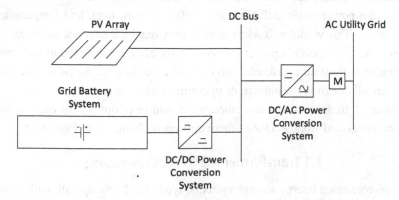

Figure 3.4 "DC-connected" grid battery system

However, there are some considerations which must be made while designing such a system:

Potential-Induced Degradation (PID) Concerns

Some, but not all PV array panels suffer from accelerated degradation, called potential-induced degradation (PID) [14], if the metal frames to which they are mounted are not held to the most negative dc rail. Therefore, many PV installations, particularly in the USA, ground the negative dc rail of the dc bus to earth. The problem with this, is that most grid energy storage systems, for safety reasons, are floating with respect to ground. As we will see later, grounding a large battery system can have serious safety drawbacks. The dc/dc conversion between an ungrounded energy storage system and a grounded PV array must be galvanically isolated. While this is certainly possible using integrated, high-frequency transformers, it will incur additional up-front costs and long-term operating losses over a non-isolated design.

Dark Matters

Most PV arrays do not like to be exposed to voltage on their output terminal when the sun is not shining. Therefore, a switch is usually placed inside the PV PCS and is opened at night. In the dc-connected storage scheme, this switch must be positioned between the PV array and the common dc bus, to allow the dc bus to carry the battery voltage during night-time operations.

Maximum Power Point

PV arrays generate the most power at an optimal voltage depending on varying conditions. Therefore, solar PCSs are designed with the ability to track the maximum power point (MPPT) under different solar, load, and temperature conditions. This works well when there is only one PCS attached to the array. However, when multiple power conversion devices are connected to the same dc bus, they must all coordinate with each other so that, for the collective load they are all drawing, the optimal dc operating voltage is set.

Given all these considerations, there are a number of options for configuring the dc-connected storage. One of these options is illustrated in Figure 3.5.

3.3 Transformer AC to AC Conversion

For ac-connected energy storage systems, typically, PCSs operate with an ac voltage that is somewhat lower than the lowest expected dc battery voltage. This is because the most common and cost-effective PCS architecture is based

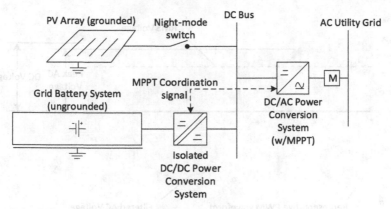

Figure 3.5 dc-connected storage with isolation, night-mode, and MPPT coordination features

on a single-stage, three-phase bridge design. In the single-stage design, the dc bus must always be higher than the highest ac voltage being converted. Multistage designs allow more flexibility in this regard, but will have increased hardware costs and reduced efficiency, and therefore are not as common.

Let us look at an example. A very common ac voltage found in many commercial buildings in the USA is 480 Vac. Because of this, many PCSs targeted for use in the USA will operate at this ac voltage. When we say the ac voltage is 480 Vac, we mean the root mean squared (RMS) value of the voltage is 480 V. The peak voltage on a 480 Vac RMS circuit is $480 \times \sqrt{2}$, or about 680 V. So, at a minimum, the dc bus voltage should never drop below this value. Additionally, manufacturers will add some margin between this value and the expected minimum dc bus voltage, of around 50 V. Therefore, the minimum dc bus voltage for a 480 Vac RMS output would be 730 Vdc.

The waveforms in Figure 3.6 show a sine wave representation of the 480 Vac voltage on the ac side of a PCS. The dc voltage is shown to be larger than the peak of the sine wave ac voltage. The representative PWM pulses show how the dc voltage is "chopped up" in varying degrees throughout the period of the sine wave. The average of these pulses would approximate the sine wave voltage. To demonstrate this concept, the diagram shows the PWM pulses at a much lower frequency than would typically be seen in a real PCS. In reality, the PWM pulses are a hundred times shorter and more frequent than shown here.

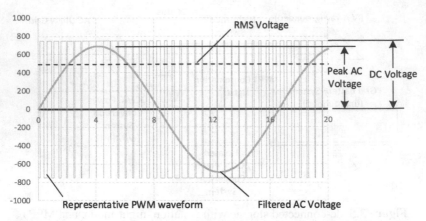

Figure 3.6 Representative waveforms of single phase PWM and filtered ac voltages

Voltage Transformation

Since the ac voltage is a function of the dc bus, among other reasons, there is often a need to transform the PCS's ac voltage to one that can readily connect to the local grid. Often, substation grid voltages are in the range of 4 kV to 15 kV. Substations are then connected to the grid through an even higher transmission voltage of 115 kV or higher. Medium-sized energy storage sites between 1 and 20 MW are typically connected to the medium voltage substation grids, while larger energy storage sites over 100 MW, are typically connected to the transmission line directly. In either case, the PCS's ac output will have to be transformed by one or more transformers (see Figure 3.7) depending on the substation design.

Safety Benefits of Isolated Transformers

Another benefit of having transformers between the grid and the energy storage is that the isolation provided by the transformers enables safer operating conditions at the energy storage site. Step-up transformers typically have a galvanic isolation barrier between one side and the other, across which energy flows through an oscillating magnetic field. Having an isolation barrier between the ground-referenced grid and the PCS, allows the PCS to have the option to "float" with respect to ground or to be ground-referenced with a hard connection.

If the PCS is referenced stiffly to ground, then whenever any part of the PCS's ac or dc power conductors inadvertently comes in contact with earth or chassis ground, there will be a significant amount of current flowing into that contact. The currents will only be limited by low-impedance power paths inside the battery or PCS

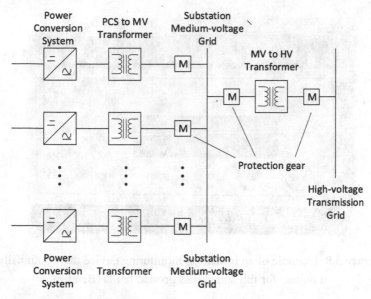

Figure 3.7 One-line diagram showing multiple stages of ac transformation between the PCSs and the grid

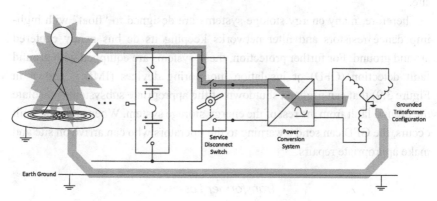

Figure 3.8 Direct current path and shock hazard through a grounded PCS

conversion circuits. If a person's body bridges a power conductor to ground as in Figure 3.8, serious injury or death can ensue.

On the other hand, if the PCS is softly referenced to ground through parasitic capacitances or high-value-sensing resistors, whenever there is a short from a dc power conductor to ground, the current that flows through the short is limited by high-impedance parasitic capacitance and resistances to ground distributed throughout the energy storage system. However, if a second short to ground

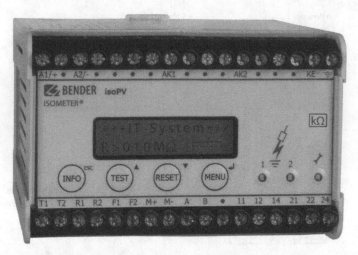

Figure 3.9 Example of an insulation monitoring device that continually monitors for full and partial ground faults (Bender)

were to happen, there would be a large amount of current flowing between both shorts and through the ground system, potentially causing a lot of damage to the site.

Therefore, many energy storage systems are designed to "float" with high-impedance resistors and filter networks keeping its dc bus gently centered around ground. For further protection, these systems are equipped with ground fault detection (GFD) or insulation monitoring devices (IMD – shown in Figure 3.9) that can trigger a shutdown of the appropriate subsystems to isolate the ground fault from the rest of the energy storage system. When an initial fault occurs, the GFD can send a warning to the operators who can arrive on site and make appropriate repairs.

Transformer Losses

There are a few losses to consider when sizing a PCS in any energy storage site with transformers and transmission lines (depicted in Figure 3.10). Transformers especially, add impedance between the PCS and the point of interconnect (POI). Because of this, PCSs must be rated with a power capacity that is higher than the required power at the POI to overcome the effects of these components.

The battery and PCS can be modeled as a voltage source and the transformer can be modeled, for the most part, by two main components, resistance and inductance (neglecting core losses for the moment), as shown in Figure 3.11.

The complex power, **S**, at the PCS will be mathematically modeled as follows:

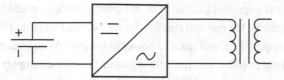

Figure 3.10 Simplified circuit diagram of the battery, PCS, and transformer

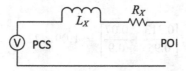

Figure 3.11 Simplified circuit model of a PCS and transformer

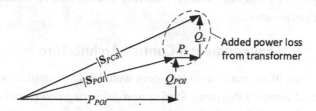

Figure 3.12 Vector diagram of the complex power losses between the PCS output and the POI

$$S_{PCS} = S_{POI} + |I_{POI}|^2 \times Z_X \qquad (3.1)$$

Graphically, this can be shown as in Figure 3.12.

Where P_x and Q_x are the real and reactive losses inside the transformer.

In addition, since the PCS powers the system's auxiliary power and the transformer's core magnetization losses while discharging, as shown in Figure 1.2, the real power output must include these extra components: $P_{aux} + P_{core}$. The total complex power provided by the PCS under full power operation will be:

$$S_{PCS} = P_{POI} \times \left(1 + \frac{R_X}{PF^2}\right) + P_{Aux} + P_{Core}$$
$$+ j\left[\frac{P_{POI}}{PF} \times \sin[\,\cos^{-1}(PF)] + P_{POI} \times \frac{L_X}{PF^2}\right], \qquad (3.2)$$

where S_{PCS} is the complex power at the PCS terminals, P_{POI} is the real power at the POI, R_x is the resistance of the transformer, PF is the power factor at the POI, and L_x is the series reactive losses in the transformer.

Therefore, it is important that the PCS and energy storage system be sized large enough to overcome the real and reactive losses between them and the POI while supplying the required real and reactive power to the grid. As an example, if the grid load is maxed out at 100%, and has a 95% power factor, the complex power at the POI will be approximately 105% of rated real system power. However, if a transformer with 1% per-unit real losses, and 7% reactive impedance is inserted between the PCS and the POI, the complex power at the PCS can be calculated to be:

$$\mathbf{S}_{PCS} = 1 + \frac{0.01}{0.9} + j\left[\frac{0.313}{0.95} + \frac{0.07}{0.9}\right] = 1.09 \angle 22°. \tag{3.3}$$

In this example, the required power at the PCS is 9% larger than the real power delivered at the POI, thus emphasizing the importance of recognizing the real and reactive losses of a transformer when considering how to size a PCS that meets the application's requirements.

4 Environmental Control Architecture

The architecture of an energy storage system would not be complete without its environmental control subsystem. Storing and delivering power and energy is only part of the total solution. Batteries, in particular, are quite finicky about their environment. Temperature and humidity are critical environmental factors in the optimization of their performance and longevity. Each battery is different, but most types enjoy a moderately cool and dry environment. Many batteries, such as lead-acid, lithium-ion, or nickel- and zinc-based batteries perform best at temperatures around 25°C, coincidentally like that in which humans thrive. Some batteries require a very tight range around this nominal value to maximize their service life. Additionally, there are high-temperature batteries which operate at hundreds of degrees and require heating equipment to keep them in their operating "sweet-spot," such as sodium sulfur.

Batteries are often housed in an enclosure, such as that shown in Figure 4.1, or building which controls the environment to keep the operating temperature in a tight range around the optimum values, the humidity dry enough to prevent condensation, and dust eliminated from the circulating air. State-of-the-art systems have airflow optimized to distribute conditioned air evenly to every battery so that the temperature differential among batteries is kept to a minimum.

A successful energy storage system will have a well-thought-out environmental control subsystem, which not only maintains the batteries and PCS at

Figure 4.1 Operating 4 MWh grid battery system, shown with cooling
infrastructure on top (Source: NEC Energy Solutions)

their optimal temperatures, but also minimizes electrical power losses and
operates reliably.

4.1 Cooling Requirements

Energy storage systems connected to the grid can be expected to handle
large amounts of power and energy. For any system that is less than perfect,
which is every system on this planet, there will be a certain amount of heat
lost in this process. This heat must be removed, or it will contribute to
a potentially harmful temperature rise of the operating components. For
example, a container of batteries described in Section 2.2.2, may have
upwards of 2 to 5 MWh of energy in them. Even if they are 95 to 98% round-
trip efficient, there will be hundreds of kilowatt-hours of heat created inside
each box every time a full cycle is executed. As such, it is important to know
the duty profile of the application being served. In some applications, where
the batteries are executing high-power discharges and charges on
a continuous basis performing frequency regulation, it is not uncommon to
see anywhere from 20 to 40 tons of air-conditioning on the top of a 53-foot
battery container. In the case of a 1-hour cycle per day application, a small
5-ton air-conditioner will be adequate to cool the enclosure and maintain an
acceptable average battery temperature.

Inside the enclosure, the cooling air must be distributed to all the heat-
generating components, so that they all experience the same operating
temperature. This is important for several reasons. If some components

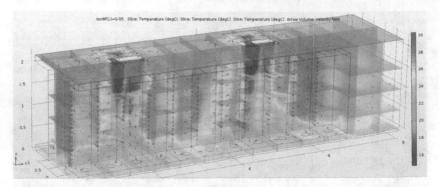

Figure 4.2 Graphical output of a thermal analysis of the inside of a battery enclosure during operation (Source: NEC Energy Solutions)

are stressed more than others, they will wear out faster. When they wear out, the rest of the system will have to work harder to make up for the lost capability, stressing the rest of the system. Thus, in effect, the whole system degrades faster than if all components are equally stressed.

Balancing the airflow to every battery component is not easy to do. However, there are computational tools that perform multiphysics, finite element analysis on physical environmental systems to show where hotspots may reside (see Figure 4.2) and allow a system designer to mitigate against them before an expensive prototype is constructed.

Temperature control at every level of the system is important. Not only do the racks and modules need to be equally cooled, but each individual cell in every module must be cooled as much as all the others. This is often not possible (see Figure 4.3) because of physical limitations built into the mechanical construction of the battery modules. In such cases, it will be necessary to match the operation of the storage system with the performance of the cooling systems.

The system should be rated to operate such that a predetermined absolute maximum and difference in temperatures are never exceeded. If, however, during normal operation, the monitoring circuits detect that the temperature limits are exceeded, it can dial back its operating power and/or send a warning message to human operators. In summary, the system should, first, be designed to operate safely under the expected operating profiles, and second, the operating profile should be dialed back if the monitored conditions exceed safe limits.

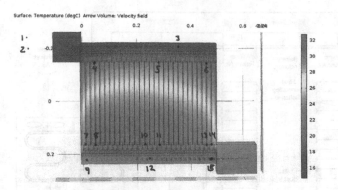

Figure 4.3 Computer simulation of a battery module with a temperature gradient from one side to another during heavy operation (Source: NEC Energy Solutions)

4.1.1 Cooling Systems

There are a few ways to remove heat from the batteries and their enclosure. The simplest and *least* recommended would be to vent the hot air from the enclosure, directly outside. Although electrically efficient and simple, it has the highest risk of introducing dust, humidity, and other contaminants into the battery environment. In fact, this method was cited as one of the reasons for some of the battery fires in South Korea in 2018 and 2019 [15]. Alternatively, air in the enclosure can be cooled with a high-volume air conditioning (HVAC) system. This can be done with an external, packaged HVAC unit, or with a split HVAC system as shown in the simplified diagrams of Figure 4.4 and Figure 4.5, respectively.

The advantage of the split system is that there are fewer ducts moving massive amounts of air around. Not only do the ducts take up space in and around the enclosure, the effort to install them contributes to considerable on-site installation costs. In both HVAC system configurations, it is crucial to design the air-distribution system such that cool air is evenly spread to each of the batteries. Computer modeling can be helpful in simulating the velocity and temperature gradients of a working HVAC system.

Finally, another approach requires tight integration with the design of the battery modules. In this approach, chilled water is distributed throughout the enclosure and into cold plates that touch each of the batteries as shown in Figure 4.6.

The advantage of this approach is that the batteries can be placed in closer proximity to each other, which reduces the number and/or size of the battery enclosures and therefore, the overall footprint of the storage

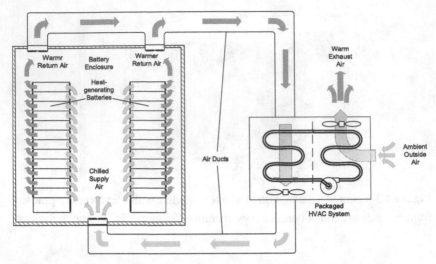

Figure 4.4 Packaged HVAC system cooling the air around the batteries in the enclosure

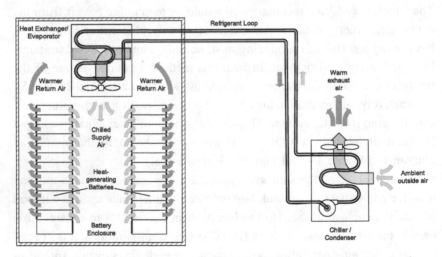

Figure 4.5 Split HVAC system cooling the air around the batteries in the enclosure

facility. Another important benefit is that it facilitates a more even temperature distribution among the batteries. With the HVAC systems, a tremendous volume of air is required to blow over the surfaces of the batteries to maintain the same amount of cooling that can be accomplished with a water loop. These benefits require additional hardware and

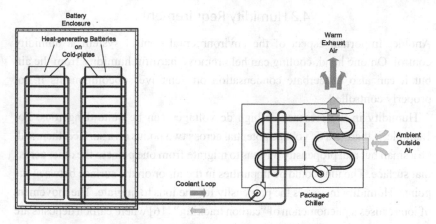

Figure 4.6 Batteries cooled directly by chilled water loop

manufacturing labor, but in some cases, may yield a larger cost savings in system balance-of-plant, installation, and battery longevity. In addition, care must be taken to ensure that the pipes do not induce condensation in close proximity to the battery cells.

4.1.2 Cooling System's Efficiency and Losses

Any cooling system, whether it is blown air or compressor-based HVAC systems, will use a certain amount of electricity to move excess heat from one place (inside the batteries) to another (outside the enclosure). A general rule of thumb is that the electrical energy required to move the heat is approximately 1/3 of the heat energy being moved.

E_E (Electrical energy required to move E_H of heat)

$= E_H$(Heat energy being moved) / 3.   (4.1)

For example, when a battery loses 10% of its energy while performing a full cycle, system designers must add another 3.3% to that figure to account for the HVAC electrical power required to remove the 10% lost in heat.

Naturally, some HVAC systems are more efficient than others, and environmental factors can play a role in affecting their efficiency. Therefore, it is important for system engineers to take into account the efficiency of the batteries, the heat generated by other auxiliary loads in the enclosure, the efficiency of the HVAC equipment and all of the environmental conditions when sizing and rating the environmental control systems of an energy storage facility.

4.2 Humidity Requirements

Another important aspect of the environmental control system is humidity control. On one hand, cooling can help remove harmful humidity from the air, but it can also exacerbate condensation on sensitive subcomponents if not properly controlled.

Humidity in the presence of high dc voltages can be quite dangerous. For example, if there is a dc voltage potential across two points on a surface, there will be a small but real propensity for ions to migrate from one point to the other across that surface. The ions could be impurities in the air or on the surface between the points. Humidity increases the propensity for the ions to migrate. The movement of ions causes a phenomenon of "carbon tracking" [16] where carbon deposits are left in the trails of the ions. The more the carbon builds up, the more conductive the tracks become, drawing more and more ions, and even electrons, across the surface. As more current flows, more heat is generated, which builds up more carbon, which increases the flow of current and adds more heat, and so on. Eventually, there will be a flashover, a fire, and maybe even an arc-flash.

Even the insulation on wires has reduced voltage withstand ratings in the presence of water. The insulation jacket on wires will often specify a lower maximum operating voltage for wet applications versus dry. The jacket in Figure 4.7 indicates 600 V for wet applications and 1,000 V for dry applications.

Needless to say, it is very important to control the humidity in a high voltage dc environment. In some environments, supplemental humidity-removing devices should be installed in the enclosure to control it.

Figure 4.7 Wires with "0.6/1kV" max operating voltages for wet and dry applications, respectively

Figure 4.8 Humidity condensing on the inside surface of a battery enclosure

Cooling the air inside a battery enclosure can help remove excessive humidity, but it is important to manage the air quality as well. Ideally, a semi-closed environment should be maintained so cool dry air recirculates around the enclosure, without introducing more humidity into the airstream. Not only is removing humidity energy intensive, but a continuous supply of external air also brings with it an unhealthy supply of dust, contaminants, fungus spores, and even insects, all of which can get into the crevices and surfaces of electronic components to reduce their reliability over a long service life. Furthermore, if humid air is introduced onto cold metal surfaces, condensation can occur (see Figure 4.8), causing unsightly corrosion on its surface at best, but even worse, water to drip onto sensitive equipment.

4.3 Heating Control

Heating is also important. Even though continual normal operation of batteries will produce an excess amount of heat that has to be removed, some situations require an addition of heat to bring the equipment into its preferred operating range. PCSs, in particular, are less reliable when they start operating in a cold environment. The thermal gradients introduced when they start up cold, induce physical stresses inside the electronic components, leading to premature failure. Batteries can also suffer accelerated capacity degradation when operated in cold conditions. For example, in lithium-ion cells, lithium-ions are less mobile through a cold electrolyte than through a warm one. This reduction in mobility causes the ions to bunch together in high concentrations, leading to lithium metal plating on the electrode surfaces. This plating can lead to gas generation, reduction in capacity, high cell impedance, and even internal short circuits in severe cases. Therefore, when a system is left unused in a cold environment for an extended period, it should be gradually warmed to its preferred operating temperature prior to use.

Fortunately, adding heat to a battery enclosure is simpler than removing it. HVAC systems commonly are available with built-in heaters, which can be used to heat the air passing through the ducts. If this is not an option, a small electric heater can be installed inside the insulated battery enclosure. Once the battery is operating, its own internal losses generally keep it at an elevated temperature, especially if the enclosure is insulated.

5 Energy Management

The last, but no less important subcomponent of energy storage described in this Element, is the management and controls system. It is often considered the "glue" that binds the other components together in a cohesive, functional, and reliable asset. In the beginning of this Element, we said, "an ideal architecture will provide the maximum benefit to the customer's grid, while maintaining the highest availability, safety, and minimum amount of lifetime cost for its operators."

A well-designed energy management system maximizes the benefits for its owner by monitoring the ESS subsystems and the grid conditions and controlling the subsystems to operate according to a pre-programmed plan. It monitors the status and condition of the battery cells and their environment. It controls the environmental conditioning systems and power conversion equipment to make sure that the energy storage is used well within its limits to avoid any unnecessary performance degradation. It protects the energy storage components by monitoring for unsafe conditions and isolating segments of it so that the remaining segments can continue to operate and service the customer's needs. Finally, it connects the customer's needs with that of the energy storage system to operate it in the most optimal manner.

5.1 Managing Energy Storage

The management of energy storage is usually performed at multiple layers. At the lowest layer, electronic circuits monitor the conditions of the individual storage cells. For batteries, this means monitoring voltage, current, and temperature of either all, or a representative number of individual cells. We introduced this low-level management in Section 2.2.1. As more and more batteries are combined to form a larger system, the monitoring circuits need to communicate with each other or with higher levels in the hierarchy to aggregate relevant control information to the site controller.

- More specifically, in a battery architecture, there may be multiple modules connected in a single rack.
- The multiple module BMSs talk to a rack BMS, which aggregates the data from each of their modules.
- Racks are often connected in a cluster of racks, (sometimes called a "zone"). The zone controller aggregates information from all its racks.
- Zones are commonly connected to a PCS, sometimes by themselves, or sometimes in parallel with one or two other zones.
- A site controller coordinates the information flow between one or more zones and the customer's controls systems and interfaces. In addition, there are many other subsystems that sense, communicate, and receive controlling commands from the site controller. These include transformers, circuit breakers, power meters, fire suppression, security, power supplies, and network systems. Figure 5.1 shows an example of such a control network from one integrator.

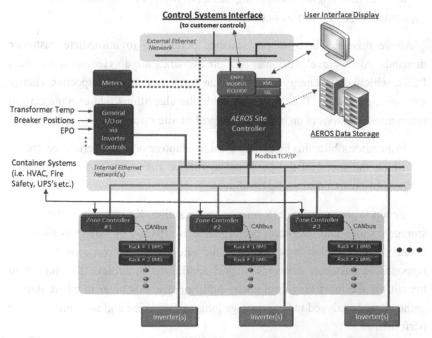

Figure 5.1 Example of a multilayer energy storage management system (Source: NEC Energy Solutions)

Figure 5.2 Three tiers of energy storage management

Conceptually, the energy storage control functions fall into several tiered categories as depicted in Figure 5.2.

At the base tier, the individual components are monitored and controlled in real time and protected from operating in a way that is unsafe for the equipment and its operators. Module BMS, rack BMS, and zone controllers all act in this layer.

⇨ Analogously, in an automobile, this lower level will include such subsystems as the engine monitoring sensors, tire-pressure monitoring, brake activators, and throttle controls.

Above this is a layer of performance pertaining to immediate customer demands. At this level, customer and site demands are divvied up to the site's PCSs, which draw energy at appropriate rates from their respective energy storage components. This layer also holds the algorithms that execute certain autonomous behaviors on behalf of the grid or site operators.

⇨ In an automobile, this layer may include adaptive or simple cruise control in which the speed of the car is set to follow a driver-specified value, and possibly respond to traffic conditions.

The top layer executes performance optimization. Knowing how the energy storage's short- and long-term performance is affected by how it is used and under what condition it is operated, the site controller can temper how it responds to customer demands or grid conditions to achieve the maximum benefit for the least long-term costs. Additionally, this layer is where data is gathered and analyzed to help manage long-term service and to improve future performance.

⇨ In the automotive analogy, the optimization layer is where the driver (whether human or not), will guide the vehicle on the most optimal path from point A to B, through traffic, and in such a way that reduces the total cost of travel.

5.2 Managing Power Conversion

As introduced in Section 3.1, PCSs have their own internal low-level control that translates external commands into signals for appropriately controlling internal switching devices. In an ESS, a site controller would be responsible for sending the control commands to the PCS to tell it what to do.

PCSs can be commanded to send energy between the energy storage and the grid in either direction. They might also be controlled to delay or advance their oscillating ac current with respect to the grid's voltage to inject or absorb reactive power into the grid. Grid operators can use these capabilities to control real power flow, regulate frequency, and influence local line voltages on their electrical grid. Therefore, to make full use of the installed energy storage, the site's energy storage controller must be able to coordinate the full capabilities of the PCS with the grid operator's requirements.

PCSs can be configured to operate in a grid-*following* mode, or a grid-*forming* mode. In the *following* mode, the PCS injects current into a grid using the grid's oscillating voltage as a timing reference for how it synchronizes the injection of the oscillating current. This is useful for when the grid operator simply wants to inject specified amounts of real and reactive power into the grid. In this mode, the grid operator could request a certain amount of power to be applied to the grid. The site controller would then send a command to the PCS to inject that power into the grid. The PCS, having already been connected and synchronized to the grid, would inject the appropriate amount of current to the grid such that the product of that current and the grid's voltage results in the desired power injected.

In the *forming* mode, the PCS follows its own internal ac voltage reference and injects current into the grid to achieve a voltage matching the magnitude and frequency of this reference. This mode is useful in situations where the grid operator requires the PCS to power an isolated (a.k.a. "islanded") microgrid as shown in Figure 5.3.

In this case, the PCS is creating, or "forming," the voltage and frequency on the islanded grid which powers the loads that depend on the grid's distribution of power. There may also be other PCSs and generators that co-generate with each other, on this grid. Tight coordination of each of the microgrid's generators is required to maintain its stability.

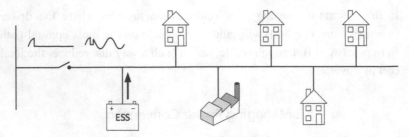

Figure 5.3 Energy storage acting in a microgrid application

5.3 Managing for Safety

When it comes to large energy storage projects, safety is of paramount importance. Not only can cells of certain chemistries, notably lithium-ion, react violently to abusive conditions, but in any large energy storage project, there is a lot of chemical and electrical energy in a relatively small space. This combination can result in significant danger to operators and equipment under the right conditions. Therefore, a multilayer approach to safety is necessary to reduce the probability of an event and to mitigate the effects of a catastrophic failure on people and the environment.

At the most basic level, the conditions of the energy storage components must be monitored to insure they never exceed safety limitations. It is also highly recommended that the sensors at this level are redundant to avoid a situation where a single point of failure can result in a failure to observe an unsafe condition.

At the next level are subsystems that can prevent the energy storage system from being exposed to dangerous conditions. This could consist of monitoring circuits that report any dangerous condition to other subsystems which can affect the conditions of the energy storage components. It could also include control circuits, which disengage and isolate the faulty energy storage component from the rest of the system. It is important to have such autonomous protection control at the lowest practical level so that a failed communication link in the higher levels of the system hierarchy does not prevent this essential protection.

Further up the hierarchy are systems that affect the power passing in and out of the energy storage system for grid functionality. In addition to conforming to the limits that are being communicated to them by the energy storage components, the power control systems keep the batteries operating well within the warranty condition limits. If these limits are maintained, the energy storage will have a greater chance to live a long, productive service life without any safety incident.

Finally, the environmental control systems need to be, first, sized appropriately for the intended customer application and environment. If under-sized, the energy storage system will experience temperature conditions outside its safe operating limits, or its performance will be severely curtailed. Second, the environmental controls systems need to respond to the conditions of the energy storage system components, not just the environment around it. For example, the HVAC system should control its rate of heat removal based on the measured temperature of the energy storage system. That way, the feedback loop is more tightly controlled and not subject to the limitations of the energy storage's heat dissipation.

Using the multilayered approach to safety, the possibility of a dangerous condition is greatly reduced, and the effects of such a condition, if one does occur, are substantially mitigated.

6 Summary

The architecture of energy storage affects its cost, performance, reliability, availability, and even its safety.

In Sections 2.1 and 2.2, we saw that choices of unitary, multielement, parallel, and series architectures can offer significant cost tradeoffs with other comparative aspects of the system, namely reliability and availability. The cost of PCS hardware, as we saw in Section 3, can also vary depending on the system's architecture choices. Higher-performing, more flexible PCS topologies will cost more up front and in ongoing services. HVAC architectures, as we saw in Section 4, can also offer tradeoffs between performance, cost, and reliability. Finally, as we saw in Section 5, the architecture of control systems can dramatically affect safety, reliability, functionality, service life, and cost.

Many times, the energy storage technology itself determines the basic architecture of the energy storage. Nevertheless, smart choices can be made about how the subcomponents conduct power and interact with each other to dramatically improve the system's benefits for its customers.

References

[1] Firstlight, "Northfield Mountain Pumped Hydro Storage Station," [Online website]. www.firstlightpower.com/facilities/?location_id=346).

[2] PNNL, [Online website]. https://caes.pnnl.gov/

[3] Highview Power [Online website]. www.highviewpower.com/technology/.

[4] Lockheed Martin, "Energy storage" [Online article]. www.lockheedmartin.com/en-us/capabilities/energy/energy-storage.html.

[5] Malta [Online website]. x.company/projects/malta/.

[6] Energy Vault [Online website]. energyvault.com/.

[7] Samsung SDI, "Energy storage system technology" [Online article]. www.samsungsdi.com/ess/energy-storage-system-technology.html.

[8] Primus Power, "Energy storage" [Online article]. www.primuspower.com/en/energy-storage/.

[9] Amber Kinetics, "West Boylston Municipal Light and Power" [Online article]. www.amberkinetics.com/portfolio/west-boylston-municipal-light-and-power/.

[10] US DOT, "U.S. Hazardous Materials Regulations 49 CFR § 173.185 Lithium batteries and cells" [Online article]. www.ecfr.gov/cgi-bin/retrieveECFR?gp=1&ty=HTML&h=L&mc=true&=PART&n=pt49.1.173#se49.2.173_1185.

[11] International Electrotechnical Commission, "Definition of a monobloc battery" [Online article]. www.electropedia.org/iev/iev.nsf/display?openform&ievref=482-02-17.

[12] D. Ouyang, M. Chen, J. Liu, R. Wei, J. Weng, and J. Wang, "Investigation of a commercial lithium-ion battery under overcharge/over-discharge failure conditions," *Royal Society of Chemistry*, vol. 8, 2018, pp. 33414–33424.

[13] Q. Wang, P. Ping, X. Zhao, G. Chu, J. Sun, and C. Chen, "Thermal runaway caused fire and explosion of lithium-ion battery," *Journal of Power Sources*, vol. 208, 2012, pp. 210–224.

[14] W. Luo, Y. S. Khoo, P. Hacke, et al., "Potential-induced degradation in photovoltaic modules: a critical review," *Energy & Environmental Science*, vol. 10, 2017, p. 43.

[15] Motie News [Online]. www.motie.go.kr/motie/ne/ps/motienews/bbs/bbsView.do?bbs_cd_n=2&bbs_seq_n=155116913.

[16] R. R. Stephenson, "Physics of DC carbon tracking of plastic," in *Fire and Materials Conference* (Motor Vehicles Fire Research Institute, San Francisco, 2005).

Acknowledgments

Thanks go to Brianna Hoff, Nadim Kanan, Roger Lin, Bill Donahue, Benjamin Schenkman and especially Diane Hoff for their guidance, support, and tedious editing.

Thanks also to the NEC Energy Solutions for the experience of building many successful energy storage projects and learning along the way.

Thanks to the contributors of photos and drawings throughout this Element, namely:

A123 Systems, LLC.
Bender GmbH & Co.
Delta Electronics, Inc.
EPC Power Corporation
LG Chem
NEC Energy Solutions, Inc.
Outback Power, Inc.
Power Sonic Corporation
Saft
Samsung SDI Co., Ltd.
Simpliphi Power, Inc.
UniEnergy Technologies

Cambridge Elements ≡

Grid Energy Storage

Babu Chalamala
Sandia National Laboratories

Dr. Babu Chalamala is manager of the Energy Storage Technologies and Systems Department at Sandia National Laboratories. He received his Ph.D. degree in physics from the University of North Texas and has extensive corporate and start-up experience spanning several years. He is an IEEE Fellow and chair of the IEEE Energy Storage and Stationary Battery Committee.

Vincent Sprenkle
Pacific Northwest National Laboratory

Dr. Vincent Sprenkle is chief scientist at Pacific Northwest National Laboratory (PNNL), and program manager for the Department of Energy's Electricity Energy Storage Program at PNNL. His work focuses on electrochemical energy storage technologies to enable renewable integration and improve grid support. He has a Ph.D. from the University of Missouri and holds 25 patents on fuel cells and batteries.

Imre Gyuk
US Department of Energy

Dr. Imre Gyuk is Director of the Energy Storage Research Program at DOE's Office of Electricity. For the past 2 decades, he has directed work on a wide portfolio of storage technologies for a broad spectrum of applications. He has a Ph.D. from Purdue University. His work has won prestigious awards including 12 R&D 100 Awards, the Phil Symons Award from ESA, and a Lifetime Achievement Award from NAATBatt.

Ralph D. Masiello
Quanta Technology

Dr. Ralph D. Masiello is a senior advisor at Quanta Technology, and developed smart grid roadmaps for several US independent system operators and the California Energy Commission. With a Ph.D. from MIT in electrical engineering, he is a Life Fellow of the IEEE, member of the US National Academy of Engineering, and won the 2009 IEEE Power Engineering Concordia Award.

Raymond Byrne
Sandia National Laboratories

Dr. Raymond Byrne is the manager of the Power Electronics and Energy Conversion Systems Department at Sandia National Laboratories, and works on optimal control of energy storage to maximize grid integration of renewables. He is an IEEE Fellow and recipient of the IEEE Millennium Medal.

About the Series

This new Elements series is perfect for practicing engineers who need to incorporate grid energy storage into their electricity infrastructure and seek comprehensive technical details about all aspects of grid energy storage. The addressed topics will span from energy storage materials to the engineering of energy storage systems. Cumulatively, the Elements series will cover energy storage technologies, distributed energy storage systems, power electronics and control systems for grid and off-grid storage, the application of stationary energy storage systems for improving grid stability and reliability, and the integration of energy storage in electricity infrastructure. This series is co-published in collaboration with the Materials Research Society.

MATERIALS RESEARCH SOCIETY®
Advancing materials. Improving the quality of life.

Cambridge Elements ☰

Grid Energy Storage

Elements in the Series

Beyond Li-ion Batteries for Grid-Scale Energy Storage
Garrett P. Wheeler, Lei Wang and Amy C. Marschilok

Energy Storage Applications in Transmission and Distribution Grids
Hisham Othman

Energy Storage Architecture
C. Michael Hoff

A full series listing is available at: www.cambridge.org/EGES

Printed in the United States
by Baker & Taylor Publisher Services

Printed in the United States
by Baker & Taylor Publisher Services